YOUR KNOWLEDGE HAS VALUE

- We will publish your bachelor's and
  master's thesis, essays and papers

- Your own eBook and book -
  sold worldwide in all relevant shops

- Earn money with each sale

Upload your text at www.GRIN.com
and publish for free

Herbert Abigaba

Internationalization of the petroleum industry. Law and policy of energy and resources in Uganda's Host Government Contracting

GRIN Publishing

Imprint:

Copyright © 2014 GRIN Verlag GmbH
Print and binding: Books on Demand GmbH, Norderstedt Germany
ISBN: 978-3-656-83515-8

This book at GRIN:

http://www.grin.com/en/e-book/280812/internationalization-of-the-petroleum-
industry-law-and-policy-of-energy

Internationalization of the petroleum industry as reflected within Uganda's Host Government Contracting

Author's Name: Herbert Abigaba*

* **Herbert Abigaba** attained a Bachelor of Science in Electrical Engineering, from the University of Dar Es Salaam, Tanzania in 2007; and a Cisco Certified Computer Networks Certificate in 2009. He wrote this paper as a coursework assignment for his Law of Energy and Resources course during the first year of his Masters of Science in Energy and Resources Management and Policy, at the School of Energy and Resources Management, Department of Engineering, University College of London, Australia. Presently in his second year (due to graduate in early November, 2014), he is a research fellow with BHP Billiton looking at energy use optimization and management strategies for Petroleum and Mining Operations, to improve cost efficiencies. He can be contacted at: herbert.abigaba.12@ucl.ac.uk

1. Introduction

This paper discusses the internationalization of the petroleum industry as witnessed within Uganda's petroleum legal and regulatory framework, drawing heavily on the Host Government Contracts (HGC). The country is selected for a number of reasons. First, the framework relates strongly with the other countries across East Africa, a region that is profoundly becoming a new petroleum frontier.[1] Second, Uganda's commercial oil discoveries realized in early 2006 have since increased to become the largest on-shore oil reserves in the Sub-Sahara Africa (SSA) over the last 20 years.[2] Third, this 'black gold' potential has seen strong foreign interest, especially over the last ten years with the country now pitted to become a top-50 oil producer in the mid-term if the development plan is achieved.[3] As a result, the country has witnessed legal, regulatory and policy reforms; strongly over the last ten years, aimed at improved management of the nascent industry as shall be discussed.

The paper is structured as follows. In the first section, a chronological development of country's petroleum legal, regulatory and policy framework is discussed. Section two explains the HGC that has been employed in Uganda to manage its hydrocarbon resources. In the third section, internationalization of the industry as reflected in Uganda's framework is considered. Although the focus is on host government contracting, the interdependence with the public law framework that governs it is appreciated. For this reason, the discussion permeates beyond the HGCs to include this framework. While it is adduced that internationalization within Uganda's legal system is apparent and still progressing, the paper suggests that developing countries face some specific challenges. In the last section, these challenges are assessed along with recommendations to remediate them, and entrench this narrative.

2. Legal Background to Uganda's Petroleum Sector

Preliminary petroleum exploration in Uganda dates back to the early 1920s when the protectorate Government geologist E. J Wayland, documented presence of hydrocarbons in the Albertine Graben

---

1 KPMG, 'Oil and Gas in Africa: Africa's Reserves, Potential and Prospects', (2013), p.18

2 B. Shepherd, 'Oil in Uganda: International Lessons for Success', (London: The Royal Institute of International Affairs, Chatham House, 2013), p.2

3 KPMG, 'Oil and Gas in Africa: Africa's Reserves, Potential and Prospects', (2013), p.16

Ð the present day sole proven hydrocarbon province.[4] Intermittent activities proceeded through to the early 1940s, when work was halted in the wake of World War II and its aftermath. With the precarious political environment that followed, consistent activities did not resume until the early 1980s when comprehensive aeromagnetic survey across the province was conducted.[5] This was followed by the World Bank project that embarked on exploration and further understanding of the Graben sub-surface in the late 1980s. This set of activities necessitated the enactment of the regulating legislation - the Petroleum (Exploration and Production) Act, Cap 150, of 1985 Ð laws of Uganda.[6]

While this Act laid emphasis on exploration operations, it was not comprehensive on production and other matters incidental such as decommissioning activities.[7] More so, there was no policy and regulatory framework in place to actualize the Act. At the same time the Act had some incoherencies with the Constitution. For instance, while the Act explicitly dealt with petroleum, under the Constitution, petroleum was considered a "mineral" whose definition was referred to the existing Mining Act of 1964, which excluded petroleum.[8] Notwithstanding, the Government of Uganda (GoU) signed the first ever Production Sharing Agreement (PSA) with *Petrofina from Belgium* in 1991 where the latter took over the entire Albertine Graben to perform further exploration work. However, given the underlying inadequacies, this PSA was a dead letter, as the licensee did not carry out any work until its expiry in 1993, and was not renewed.[9]

As a remedy to the regulatory and institutional framework inadequacies, the Petroleum (Exploration and Production) Regulations were passed in 1993 by the newly created Petroleum Exploration and Production Department (PEPD) within the Ministry of Energy and Mineral Development (MEMD).[9] PEPD was formed from a small petroleum unit under the Mineral and Geology Department signifying the initial institutional reform for the sector. These reforms were followed by the promulgation of a new national Constitution of 1995. Article 244 of the new Constitution was fundamental in closing the lacunas in the legal framework as it explicitly provided for minerals and petroleum. It created a fundamental shift where the applicable laws at the time were harmonized with the existing law and regulations. Over this period, limited work happened, until in 1997 when the second PSA was signed between the GoU and Heritage Oil and Gas Company from UK (Heritage Oil) for one of the nine exploration areas of the Graben.[9]

A year later, Energy Africa (now Tullow Oil) from Ireland came on board, followed by Hardman Resources from Australia in 2001. The three International Oil Companies (IOC) largely participated

---

4 'Kashambuzi, R. & Mugisha, F.; Exploration History and Petroleum Potential of Uganda', Petroleum Exploration and Production Department (PEPD), Government of Uganda (GoU), ([Accessed September, 8 2013]. Available from: http://www.energy.eac.int/eapc2003/contents/22a.pdf

5 ibid.

6 PEPD. (2013). "Petroleum Potential of the Albertine Graben." [Accessed September 26, 2013], from http://www.petroleum.go.ug/uploads/PEPD Brochure November%202008.pdf

7 Arthur Bainomugisha, and Benson Tusasirwe, 'Escaping the Oil Curse and Making Poverty History. A Review of the Oil and Gas Policy and Legal Framework for Uganda', (ACODE Policy Research, 2006), p.22---26

8 ibid.

9 PEPD, 'The History and Progress of Petroleum Exploration and Development in Uganda.'2014) <http://www.petroleum.go.ug/page.php?k=abthistory > [Accessed September 24, 2013].

in the sector until 2005 when another IOC - Tower Resources (Neptune) from UK joined the pack.[9] The heightened activity and increasing number of IOCs necessitated yet another review of legal and regulatory framework to streamline the management of the sector. As a result, the petroleum Act 1985 was revised in 2000; followed by the formulation of the Energy Policy in 2002, and the Petroleum Supply Act, 2003 that was to govern downstream activities.[10]

The key objectives of these reforms were to improve monitoring and regulation (especially aspects like work programs and environmental protection) as well as provide for a conducive environment for investment into the sector.[10] Consequently, a model PSA was developed in 2000, and will be considered in detail later.[11] However, due to the increased role of IOCs over the following years, the reforms proved still inadequate. For instance, they remained more comprehensive for exploration, but less for production and other downstream aspects. Nonetheless, this framework was crucial in driving the activities of the sector until major commercial oil discoveries in 2006.

2.1 Post‑‑‑2006 Era

The realism of commercial oil discoveries coupled with the presence of IOCs, and other international interests thus required fit-for-purpose framework to effectively manage the new challenges. This led to the formulation of the National Oil and Gas Policy (NOGP) in 2008. The NOGP provided for a comprehensive framework covering the entire value chain of the industry, including tail-end issues like site restoration and oil revenue management.[12] The policy articulates key aspects that include, *inter alia;*

- Development of legislation and expertise for effective management and regulation of the sector
- Ecologically sustainable development that includes local content participation and a prudent depletion policy
- Provision of a conducive environment for attracting the different levels of investment
- Establishment of an institutional framework separating Policy work Ð by the PEDP, from regulation Ð by the Petroleum Authority of Uganda (PAU), and business management Ð by the National Oil Company (NATOIL).
- Development of a refinery to satisfy both local and regional demand Ð partly in line with the Strategy for the Development of Regional Refineries as passed by the East African Community (EAC) Summit of Heads of State in 2008 - pursuant to Article 101 of the EAC Treaty.[13]

Due to these ideals among others, the policy has been hailed as being progressive and a standard document that incorporates numerous international "best practices and principles" such as intergeneration equity; ecologically sustainable development; transparency and local content

---

10 'Kashambuzi, R. & Mugisha, F.; Exploration History and Petroleum Potential of Uganda', Petroleum Exploration and Production Department (PEPD), Government of Uganda (GoU), ([Accessed September, 8 2013]. Available from: http://www.energy.eac.int/eapc2003/contents/22a.pdf

11 PEPD, 'Model Production Sharing Contract for Petroleum Exploration, Development & Production in Uganda', (1999)

12 'The National Oil and Gas Policy for Uganda (NOGP), February 2008', Ministry of Energy and Minerals Development (MEMD), (<http://www.acode‑‑‑u.org/documents/oildocs/oil&gas_policy.pdf> [Accessed September, 16 2013].

13 EAC, 'Strategy for the Development of Regional Refineries', (Arusha, Tanzania: East African Community Secretariat, 2008).

participation.[14] These principles are now benchmark for best practise across the international industry. Further still, pursuant to the aspects above, the Petroleum (Exploration, Development and Production) and (Refining, Transmission and Storage) Act(s) 2013 were enacted to operationalize the NOGP and in effect repealed the earlier Act, Cap 150 of 1985 [rev. 2000].[15]

The old inadequacies coupled with the new and emerging challenges were key drivers for enacting the new laws.[12] This issue cannot be overemphasized. Out-dated laws, apart from the associated management inefficiencies, are also unlikely to inspire confidence, as they do not accurately reflect the current market trends.[16] For IOCs, the confidence relates with mitigating financial and political risk. For government, a strong and dependable framework leverages strong negotiations and effective management of the above associated matters.[16] For the same reason, Uganda placed a moratorium on all new petroleum licensing in 2007 until the full actualization of the NOGP – particularly the three key laws of 2013 to manage the different aspects of the value chain.[17]

However, this moratorium was lifted earlier (in 2012), to allow Tullow Oil to proceed with the impending purchase and sale agreement that is set to open over $10 billion of investment into the sector. Under this deal, Tullow purchased Heritage Oil's stake for $1.45 billion in 2011 and then farmed-down to Total (France) and China National Off-shore Oil Corporation (CNOOC) in a deal worth nearly $3 billion a year later, with each now owning a third of Tullow Oil's tenement. The signing of these respective PSAs brought to a total of five the number of active PSAs and IOCs in Uganda's Petroleum sector. This case is a mild version of how IOCs can actually influence the host-state interests in developing countries, which are often divided between attracting investment and protecting national ideals and aspirations.

3. Uganda's Host Government Contract

HGC can generally be defined as an agreement through which the host government (owner of the hydrocarbons) grants rights to a private company to explore for and develop hydrocarbons within a given area, and compensates such a company for undertaking the associated risks.[21] HGCs are governed and construed under public law, which differentiates them from private law contracts such as the ones alluded to in the previous section that are between private entities. Four major types of HGCs exist across the international industry. These include: modern concessions, production-sharing agreements (PSA), joint venture agreements, and service contracts. Uganda has employed the PSA type, and proceeded to develop a model to guide the relevant participants. Under its PSA, when successful, the licensee pays a royalty on gross production, and then deducts a pre-specified share of the gross production to recover capital, development and production costs [cost oil]. The

---

14 Izama, A. and T. Otoa, 'Status of Uganda's Oil and Gas Legislation', (ACODE Policy Research, Kampala, Uganda, 2011), p. 3---4

15 The Uganda Government: Petroleum (Exploration, Development and Production) Act, 2013', in *CVI* (Entebbe, Uganda: UPPC), Available from:https:// http://www.petroleum.go.ug/page.php?k=regacts

16 Risks Control, 'A New Frontier: Oil and Gas in East Africa', (Nairobi: 2012), p.4

17 See event dated 2012, February 3. http://www.oilinuganda.org/categories/oil---timeline

reminder [profit oil] is then shared between the GoU and the licensee as stipulated in the agreement, and the licensee pays agreed taxes off its share of the profit oil.[18]

Importantly, a review of Uganda's model PSA suggests a corpus of many international industry standard clauses that would be found within model contracts developed by supranational associations such as Association of International Petroleum Negotiators (AIPN).[19] Notwithstanding, the NOGP highlighted inadequacies with some of these clauses such as those pertaining the fiscal and local content terms.[12] While not possible at the moment to establish whether an improved model PSA exists, it is possible to argue that Uganda's model PSA is work in progress, and future lessons will provide even better opportunities for continuous improvement.[20]

4. Internationalization of Uganda's Legal Spectrum

International practices and principles have increasingly become embedded into the PSA and their governing law over last two decades.[21] These relate to, but are not limited to, the use of model contracts with standard clauses as well as convergence of other aspects pertaining the legal, regulatory and policy framework. Globalization, transfer of industry acceptable practices and the role played by the concentrated number of major IOCs have been both cause-changers as well as indicators of this convergence. This section discusses these issues with respect to Uganda.

4.1 The Use of Stabilization Clauses (SCs)

A stabilization clause is a means by which foreign companies mitigate risks associated with their investments. It addresses how changes to the law following the execution of the HGC can be treated, and to what extent such changes affect the rights and obligations of parties.[22] While they have been used historically to mitigate political risks associated with expropriation and instability, they have since evolved to cover other risks such as currency inconvertibility and tax law changes.[23] Generally SCs are today categorized into two: the traditional freezing type and the modern-hybrid type. The former clause provide that the law in effect when the contract was executed governs the contract throughout its lifespan or freezes the legislation governing the contract, while the latter aims at avoiding any unilateral modification of the contract, and particularly provides that any changes have to be mutually agreed upon, and the clause sets out the procedure how.[23]

18 *Production Sharing Agreement (PSA), By and Between the GoU and Heritage Oil and Gas Limited, January 2007*, Kampala, Uganda, viewed September, 3 2013,
<http://www.platformlondon.org/carbonweb/documents/uganda/leaked Uganda PSA Block3A Heritage part1.pdf%3E.

19 AIPN, 'AIPN Model Contract International Dispute Resolution Agreement', (2004)

20 IIED, 'How to Scrutinize a Production Sharing Agreement', (International Institute for Environment and Development (IIED), 2012), 16---17

21 ibid., p.19---24

22 A. A. Chekol, 'Stabilization Clauses in Petroleum Development Agreements: Examining Their Adequacy and Efficacy', (Dundee, Scotland University of Dundee, 2008), 4---5.

23 A. F. M. Maniruzzaman, 'The Pursuit of Stability in International Energy Investment Contracts: A Critical Appraisal of the Emerging Trends', *Journal of World Energy Law & Business,* 1 (2008), 121---57.

The modern-hybrid clause has now become prevalent within international investment contracts,[24] and is the same applied within Uganda's model PSA *(Article 31)*.

> In Heritage Oil's PSA, the modern-hybrid clause (Article 33) provides that: *If, following the Effective Date, there is any change in the laws or regulations of Uganda which materially reduces economic benefits derived or to be derived by the licensee hereunder, Licensee may notify the GoU accordingly and thereafter the Parties shall meet to negotiate in good faith and agree upon, the necessary modifications to this Agreement to restore the Licensee to substantially the same overall economic position as prevailed hereunder prior to such change(s). In the event that the Parties are unable to agree... then either Party may refer the matter for determination pursuant to paragraph 26.1 [Arbitration Clause].[18]*

While such clauses instill confidence among investors, the legal significance and implementation of SCs remains contentious.[22] This is due to two reasons. First, the clause creates a conflict of ideals and aspirations for developing countries like Uganda. On the one hand, they are perceived as beneficial in mitigating risk for foreign investors and are hence important in attracting foreign direct investment (FDI).[23] On the other, they can prevent the host-state from taking actions necessary to protect the rights of the citizens or enforcing laws similar to those regulating other sectors.[22] In Uganda's case, this was visible during the formulation of the latest laws of 2013. The national parliament had recommended inclusion of an explicit sub-clause under *Article 6* of the Petroleum Act, 2013 defining the extent to which stabilization law can be negotiated, particularly when economic equilibrium aspects are to be triggered.[25] Instead, the final law provides handling such matters based on a model contract, which is less explicit.[15]

Although the parliament proposal could have been more stringent, this is an example where the executive works on feedback of foreign influence to compromise certain positions perceived to be domestically popular.[26] However, such conflicting interests with the sovereignty and national interests could in the long run undermine the stability of SCs.[27] This then raises the view that while SCs reflects internationalization; they can also be overtaken by events where the will and framework to anchor on is inadequate.[28]

---

24 J.N. Emeka, 'Anchoring Stabilization Clauses in International Petroleum Contracts.' *International Lawyer,* 42 (2008), 1317--- 38

25 Parliament of Republic of Uganda, Report of the Parliamentary Committee on the Petroleum Bill, 2012 [Retrieved on September 26, 2013 from http://www.parliament.go.ug/

26 Arthur Bainomugisha, and Benson Tusasirwe, 'Escaping the Oil Curse and Making Poverty History. A Review of the Oil and Gas Policy and Legal Framework for Uganda', (ACODE Policy Research, 2006), p.24---7

27 Alexia Brunet, and Juan Agustin Lentini, 'Arbitration of International Oil, Gas, and Energy Disputes in Latin America', *Nw. J. Int'l L. & Bus,* 27 (2006), 592---611

28 A. A. Chekol, 'Stabilization Clauses in Petroleum Development Agreements: Examining Their Adequacy and Efficacy', (Dundee, Scotland University of Dundee, 2008), 7---10.

4.2 International Dispute Resolution

Dispute resolution mechanism represents different techniques through which parties may resolve disputes.[29] Common examples include the negotiation, mediation, conciliation, expert appraisal, arbitration and litigation. This section focuses on arbitration as the main method that has been applied in Uganda's PSA. Under arbitration, parties to a dispute present the facts of their case to a neutral third party for determination.[29] A typical arbitration clause postulates an arbitral venue, and where necessary the relevant treaty protection, whether Bilateral Investment Treaty (BIT) or Multilateral Investment Treaty (MIT).

> In Heritage Oil's PSA, the arbitration clause – Article 26 provides that: *Any dispute arising under the PSA and not settled amicably within sixty (60) days, shall be referred to Arbitration in accordance with the United Nations Commission for International Trade Law (UNCITRAL) rules.... The said arbitration shall take place in London; England...The Arbitration Award shall be final and binding on the Parties.*[18]

However because there is no set of uniform rules and the precedential value of tribunals is limited,[30] Uganda's model PSA (Article 23) provides for International Centre for Settlement of Investment Disputes (ICSID) Arbitration rules and UNCITRAL Arbitration rules as an alternative.[11] To the extent, this provides a set of international dispute resolution protection for IOCs in Uganda. However, interpretation of applicable laws and provisions to the PSA remains critical.[30] This scenario was evident in the most significant tax dispute experienced in the country so far - between the GoU versus Heritage Oil, over the sale of the latter's assets to Tullow Oil.[31]

While Heritage Oil argued that the dispute be settled under Article 26 (above), the host-state held that the matter was purely related to 'taxation' and hence fell within Article 14 of the PSA that provided *that all applicable taxes, levies and other impositions were to be paid in accordance with the laws of Uganda.* For this reason, the GoU referred the matter to the Uganda Tax Appeals Tribunal, which was determined in its favour.[32] An appeal by Heritage Oil to the High Court to stop the proceedings of the Tribunal was not upheld. Since Tullow Oil (the buyer) had vested interests in closing the transaction and attaining GoU consent for the impending farm-down, a Memorandum of Understanding (MOU) was signed. Under the MOU, Tullow Oil committed to pay the taxes under dispute, and take over the dispute as government's third party agent.[31] This process allowed the Parties to refer the matter for settlement under international arbitration in London in accordance with Article 26, where the court upheld the rulings and findings of the Ugandan courts.[31]

---

29 A. Timothy Martin, 'Dispute Resolution in the International Energy Sector: An Overview', *The Journal of World Energy Law & Business,* 4 (2011), 336---42.

30 Gantz, D. A. (2004), "Investor---State Arbitration Under ICSID, the ICSID Additional Facility and the UNCTAD Arbitral Rules", p.1---2

31 Izama, A & Mulangwa, HW 2011, 'Understanding the tax dispute: Heritage, Tullow, and the Government of Uganda', Advocates Coalition for Development and Environment (ACODE), Kampala.

32 Heritage Oil & Gas Ltd Vs. Uganda Revenue Authority (Civil Appeal No. 14 of 2011), retrieved on March 23, 2014 from http://www.ulii.org/ug/judgment/commercial---court/2011/97

The precedence set by the above dispute provides strong neutrality imperative for parties in managing future disputes, especially for IOCs seeking to invest billions of dollars.[33] For the GoU, this is crucial in entrenching the culture while subsiding held apprehensions.[34] This is even more critical as there is already another dispute pending before the ICSID – *see 'Tullow Uganda Operations PTY LTD v. Republic of Uganda (ICSID Case No. ARB/12/34)*. While this section meant to highlight the extent to which international arbitration has taken root within Uganda's PSAs, the details also elaborate the possible routes IOCs can take in case of such disputes arise.

4.3 The Influence of Foreign Companies and Globalization

Petroleum exploration and production (E&P) activities are picking pace across the East African region. According to a Deloitte 2013 report, more hydrocarbons have been discovered in the region over the last two years than anywhere else in the world. [35] This is based on the following factors. First, is the discovery of enormous oil reserves in Uganda, now estimated at 3.5 billion barrels with further exploration and appraisal still on-going. Secondly, neighbouring Kenya has indicated similar oil prospects. Thirdly, off the seaboard of Tanzania and Mozambique, significant natural gas strikes has been confirmed, and are projected to make the region the third-largest gas exporter in the world in the long-term.[35] As a result, foreign interest in the region's petroleum sector has been growing strongly over the years.

In Uganda's legal regime, the foreign interest has come in two dimensions. First, the role played by the IOCs in transferring international industry good practice that is progressively becoming enshrined in the governing laws. For example, Article 25 of the PSA between *GoU and Heritage Oil* as well as the environmental liability clauses of the Petroleum Act 2013 articulates such aspects as "...operations are to be carried out in a safe environmentally acceptable manner consistent with the good international industry practice and applicable lawsÉ."[18] Apprehensions about implementation and enforcement notwithstanding, this has meant an increasing sector effort towards upholding these practices. A typical case is where Tullow Oil has been compelled to act in this manner out of its own corporate charter and national legislation requirements.[36]

Perhaps, even more significant for IOCs is the recent commitment by G8 countries to put laws and regulations in place requiring oil, gas and mining companies to disclose all payments to governments under the Extractive Industries Transparency Initiative (EITI).[37] The EITI is a global coalition of governments, IOCs, civil society and institutional investors aiming to promote transparency and sustainable development. With the EU and the U.S. having since passed such "Publish What You Pay" regulations, it means all the major IOCs operating in Uganda will now

---

33 A. Timothy Martin, 'Dispute Resolution in the International Energy Sector: An Overview', *The Journal of World Energy Law & Business*, 4 (2011), 332---68.

34 Carmen Otero García---Castrillón, 'Reflections on the Law Applicable to International Oil Contracts', *The Journal of World Energy Law & Business*, 6 (2013), 129---62

35 Deloitte, 'The Deloitte Guide to Oil and Gas in East Africa Where Potential Lies', (Dar es salaam, Tanzania: Deloitte & Touche, 2013).

36 Nangendo, F. and T. Fahey, 'Economic Displacement: A case study of oil exploration in Uganda', (Tullow Oil Uganda, 2013), 2---4, available from http://www.iaia.org

37 EITI, New Disclosure Requirements for the Extractive Industry: [cited 2013 September, 12]. Available from: http://europa.eu/rapid/press---release MEMO---13---541 en.htm

make public such information.[37] Such international influences will inevitably compel Uganda to adjust and conform to such industry practices, although to which level will be underpinned by the state of governance.[38]

Second, there is the increasing convergence of the international petroleum industry.[39] Traditionally, the industry has been globalized with a concentrated number of major IOCs highly cooperating around various industry activities.[40] Over last decades, "new" globalization, privatization, regional integration et al., has meant jurisdiction legal, policy and regulatory instruments are fast becoming trans-national.[41] Such instruments are now developed following a degree of comparative work. Uganda's recent sector framework had a rich input of lessons and experiences from Africa and beyond, while Norway provided a great deal of technical support.[12] This has been exacerbated by training, research and industry literature that has become equally globalised. Ultimately, the pool available for practitioners and experts is increasingly common irrespective of the geographical location.[41] Based on this convergence; there is an industry consensus that this is fast recognizing a specific legal regime that deals with the international petroleum industry.[41]

5. Lessons, Challenges and Recommendations

From the foregoing, it is clear that many international energy law, practices and norms have found (or continue to find) their way into Uganda's petroleum governing framework. However, for East African countries, and Uganda in particular, this narrative comes against some specific challenges. Firstly, while Uganda has had significant improvements in its legal and regulatory reforms over the last decade, the 2012 country risk report suggests prevalent inadequacies related to weak institutions and political management.[42] This thus creates a situation where the improved framework does not necessarily produce anticipated outcomes. Consistent effort towards building institutional capacity and enhancing the spirit of the framework through good governance will remain crucial.[43]

Secondly, the tax dispute between Heritage Oil and the GoU signaled significant issues of concern. One, there appears to be uncertainty surrounding international dispute resolution mechanisms that is proffered by either nationalistic innuendos, politics or the relative costs involved. Public debate was skewed on whether to have the case heard under the local court or international court system. While this case was ruled in favour of the host-state, it would be interesting to imagine the future precedence for the contrary.[44] The other issue relates to Tullow Oil signing an MOU with the GoU

---

38 B. Shepherd, 'Oil in Uganda: International Lessons for Success', (London: The Royal Institute of International Affairs, Chatham House, 2013), 4---5

39 K. Talus, 'OGEL Ten Years Special Issue: Internationalization of Energy Law', Oil, Gas and Energy Law (OGEL), (2012) <http://www.ogel.org/article.asp?key=3264>.

40 UNCTAD, 'The Role of International Investment Agreements in Attracting Foreign Direct Investment to Developing Countries', in *UNCTAD Series* (New York & Geneva: UN, 2009)

41 Kim Talus, Scott Looper, and Steven Otillar, 'Lex Petrolea and the Internationalization of Petroleum Agreements: Focus on Host Government Contracts', *Journal of World Energy Law and Business*, 5 (2012), 181---93.

42 Risks Control, 'A New Frontier: Oil and Gas in East Africa', (Nairobi: 2012), p.5---8

43 . Shepherd, 'Oil in Uganda: International Lessons for Success', (London: The Royal Institute of International Affairs, Chatham House, 2013), 9---10

44 Angelo Izama, and Hashim Mulangwa, 'Understanding the Tax Dispute: Heritage, Tullow, and the Government of Uganda', (Kampala: Advocates Coalition for Development and Environment (ACODE), 2011), pp. 1---6.

to act as its tax agent, which was legally questionable.[44] As stated in the court pleadings, Heritage Oil viewed this as an act of self-enrichment on the part of Tullow Oil, given that the later had a vested interest for the impending farm-down deal that was in bad faith.[45] Hence this autopsy not only provides an insight into commercial tactics, but also regulatory bottlenecks and politics that could irretrievably affect future dealings. Insertion of formal rules into the existing framework could make for improvement in this area.[44]

Thirdly, the paucity of legal expertise combined with a lack of general knowledge about the extractive industry value chain among domestic practitioners.[21] The implication is sometimes the regulation and legal effectiveness is as good as the local expertise, which then undermines the capacity to handle contracts drafting, negotiations and execution. Embracing peer supranational associations such as AIPN and strong policies to enhance the human resource base come highly recommended under this spectrum.

Lastly, in Article 13 of Heritage Oil PSA, Cost Recovery Factors are used for production sharing based on levels of production, but also ring fenced to cost-recovery limits. These mechanisms inevitably require detailed and flexible controls to ensure government financial interests are properly protected within changing market conditions.[46] For instance, it is possible that the latest PSAs have been ameliorated to include the increasingly acceptable price-linked fiscal terms such as windfall taxes.[47] Creating a balance that guarantees protection of host-state benefits while reducing uncertainty of the tax system will remain important is sustaining investment for growth of the industry.[47]

7. Conclusion

The paper discussed the internationalization of the petroleum industry as reflected in Uganda's host government contract, basing on the indicators, causes and challenges. The GoU has initiated various reforms over the recent years aimed at the efficient governing of its nascent petroleum sector. It is quite clear the various reforms in general and the existing PSA developed over this period in particular have all benefited and adopted experiences, lessons and models from the international petroleum industry. These have been in part been driven by the historical nature of the industry, and in part by the influence of globalization and a concentrated number of IOCs; meeting with the host state's desire to attract and sustain foreign investment.

However, like in many developing countries, there are still prevalent challenges to contend with. These relate to: weak institutions, political management, and paucity of legal expertise, and pseudo uncertain tax system that balances national interests with sustained investment. While this paper draws recommendations to remediate these challenges, concerted effort and further work to address these will be required, so as to raise Uganda's sector to a strong level of international industry good practice.

---

45 Taimour Lay (2013), in the matter between Tullow Uganda versus Heritage Oil and Gas Ltd.: High Court of Justice Queen's Bench Division Commercial Court. [cited 2013 September, 12]. Available from: http://clients.squareeye.net/uploads/oec/judgments/20130614---DavidWolfson---RichardMott---TullowUgandaJudgment.pdf

46 S.P. Otillar; J.C. Sonnier, 'OGEL Special Issue: Host Government Contracts in the Upstream Oil and Gas Sector', Oil, Gas and Energy Law (OGEL), (2010) <http://www.ogel.org/article.asp?key=3047>.

47 H.A. Kallisa, 'What Amendments Could Be Made to Uganda's Current Model PSA to Achieve the Economic Objectives in the National Oil and Gas Policy for Uganda?' (University of Dundee, 2010)

8. Bibliography

1. KPMG, 'Oil and Gas in Africa: Africa's Reserves, Potential and Prospects', (2013)
2. B. Shepherd, 'Oil in Uganda: International Lessons for Success', (London: The Royal Institute of International Affairs, Chatham House, 2013)
3. 'Kashambuzi, R. & Mugisha, F.; Exploration History and Petroleum Potential of Uganda', Petroleum Exploration and Production Department (PEPD), Government of Uganda (GoU), ([Accessed September, 8 2013]. Available from: http://www.energy.eac.int/eapc2003/contents/22a.pdf
4. PEPD. (2013). "Petroleum Potential of the Albertine Graben." [Accessed September 26, 2013], from http://www.petroleum.go.ug/uploads/PEPD Brochure November%202008.pdf
5. Arthur Bainomugisha, and Benson Tusasirwe, 'Escaping the Oil Curse and Making Poverty History. A Review of the Oil and Gas Policy and Legal Framework for Uganda', (ACODE Policy Research, 2006)
6. PEPD, 'The History and Progress of Petroleum Exploration and Development in Uganda.'2014) <http://www.petroleum.go.ug/page.php?k=abthistory > [Accessed September 24, 2013].
7. PEPD, 'Model Production Sharing Contract for Petroleum Exploration, Development & Production in Uganda', (1999)
8. 'The National Oil and Gas Policy for Uganda (NOGP), February 2008', Ministry of Energy and Minerals Development (MEMD), (<http://www.acode---u.org/documents/oildocs/oil&gas_policy.pdf> [Accessed September, 16 2013]
9. EAC, 'Strategy for the Development of Regional Refineries', (Arusha, Tanzania: East African Community Secretariat, 2008).
10. Izama, A. and T. Otoa, 'Status of Uganda's Oil and Gas Legislation', (ACODE Policy Research, Kampala, Uganda, 2011)
11. The Uganda Government: Petroleum (Exploration, Development and Production) Act, 2013', in *CVI* (Entebbe, Uganda: UPPC), Available from:https://http://www.petroleum.go.ug/page.php?k=regacts
12. Risks Control, 'A New Frontier: Oil and Gas in East Africa', (Nairobi: 2012)
13. *Production Sharing Agreement (PSA), By and Between the GoU and Heritage Oil and Gas Limited, January 2007*, Kampala, Uganda, viewed September, 3 2013,
14. <http://www.platformlondon.org/carbonweb/documents/uganda/leaked Uganda PSA Block3A Heritage part1.pdf%3E.
15. AIPN, 'AIPN Model Contract International Dispute Resolution Agreement', (2004)
16. IIED, 'How to Scrutinize a Production Sharing Agreement', (International Institute for Environment and Development (IIED), 2012)
17. A. A. Chekol, 'Stabilization Clauses in Petroleum Development Agreements: Examining Their Adequacy and Efficacy', (Dundee, Scotland University of Dundee, 2008)
18. A. F. M. Maniruzzaman, 'The Pursuit of Stability in International Energy Investment Contracts: A Critical Appraisal of the Emerging Trends', *Journal of World Energy Law & Business,* 1 (2008)
19. J.N. Emeka, 'Anchoring Stabilization Clauses in International Petroleum Contracts.' *International Lawyer,* 42 (2008)
20. Parliament of Republic of Uganda, Report of the Parliamentary Committee on the Petroleum

Bill, 2012 [Retrieved on September 26, 2013 from http://www.parliament.go.ug/

21.   Alexia Brunet, and Juan Agustin Lentini, 'Arbitration of International Oil, Gas, and Energy Disputes in Latin America', *Nw. J. Int'l L. & Bus,* 27 (2006)

22.   A. Timothy Martin, 'Dispute Resolution in the International Energy Sector: An Overview', *The Journal of World Energy Law & Business,* 4 (2011)

23.   Gantz, D. A. (2004), "Investor---State Arbitration Under ICSID, the ICSID Additional Facility and the UNCTAD Arbitral Rules"

24.   Izama, A & Mulangwa, HW 2011, 'Understanding the tax dispute: Heritage, Tullow, and the Government of Uganda', Advocates Coalition for Development and Environment (ACODE), Kampala.

25.   Heritage Oil & Gas Ltd Vs. Uganda Revenue Authority (Civil Appeal No. 14 of 2011), retrieved on March 23, 2014 from http://www.ulii.org/ug/judgment/commercial---court/2011/97

26.   Carmen Otero García---Castrillón, 'Reflections on the Law Applicable to International Oil Contracts', *The Journal of World Energy Law & Business,* 6 (2013)

27.   Deloitte, 'The Deloitte Guide to Oil and Gas in East Africa Where Potential Lies', (Dar es Salaam: Deloitte & Touche, 2013).

28.   Nangendo, F. and T. Fahey, 'Economic Displacement: A case study of oil exploration in Uganda', (Tullow Oil Uganda, 2013)

29.   EITI, New Disclosure Requirements for the Extractive Industry: [cited 2013 September, 12]. Available from: http://europa.eu/rapid/press-

30.   K. Talus, 'OGEL Ten Years Special Issue: Internationalization of Energy Law', Oil, Gas and Energy Law (OGEL), (2012)

31.   UNCTAD, 'The Role of International Investment Agreements in Attracting Foreign Direct Investment to Developing Countries', in *UNCTAD Series* (New York & Geneva: UN, 2009)

32.   Kim Talus, Scott Looper, and Steven Otillar, 'Lex Petrolea and the Internationalization of Petroleum Agreements: Focus on Host Government Contracts', *Journal of World Energy Law and Business,* 5 (2012)

33.   Taimour Lay (2013), in the matter between Tullow Uganda versus Heritage Oil and Gas Ltd.: High Court of Justice Queen's Bench Division Commercial Court. [cited 2013 September, 12]. Available from: http://clients.squareeye.net/uploads/oec/judgments/20130614---DavidWolfson---RichardMott---TullowUgandaJudgment.pdf

34.   S.P. Otillar; J.C. Sonnier, 'OGEL Special Issue: Host Government Contracts in the Upstream Oil and Gas Sector', Oil, Gas and Energy Law (OGEL), (2010)

35.   H.A. Kallisa, 'What Amendments Could Be Made to Uganda's Current Model PSA to Achieve the Economic Objectives in the National Oil and Gas Policy for Uganda?' (University of Dundee, 2010)